Written by Jenny Phillips
Cover design by Kayla Ellingsworth
Illustrated by Vanessa Toye

goodandbeautiful.com

"Martha! Otto!" Mom called me and my twin brother, Otto. "Let's go on a mushroom walk."

We left our farmhouse in the Black Forest, Germany.

The morning sun slanted across the hills and valleys as we walked on a little dirt path. Mom took our hands.

"I've told you before that some mushrooms are delicious and healthy, but some are very dangerous."

“Today I am going
to start teaching you
how to know which
mushrooms are safe.”

We entered the cool, quiet forest. A little brook gurgled over mossy rocks. Mom continued talking to us.

"Today, I also want to teach you about another kind of safety—sexual abuse safety."

“What is sexual abuse?” Otto asked.

“That is when someone touches the private parts of your body or takes pictures of you naked. Things like that,” Mom explained.

“Does that really happen?” I asked.
We paused by a tree to watch a squirrel scamper around the branches.
“Yes,” said Mom, “unfortunately. The Bible teaches us how wrong it is to harm a child in any way, but sometimes people make really wrong decisions.”

"Often it can even be people you know. So, I want to teach you some important things."

I reached up for Mom's hand again.

"Thank you so much for teaching us and wanting us to be safe, Mom!"

Mom squeezed my hand. "Oh, look," she said, kneeling down. "These mushrooms are delicious and totally safe."

We helped her pick them and put them in her basket.

Mom continued talking. “Sometimes it is okay for you not to have your clothes on in front of others. For example, parents have to help younger kids dress and bathe. It’s also okay for a doctor to take a look at your private areas if your parent is with you and says it is okay.”

We stopped on a grassy bank and took off our shoes. I dipped my feet in the cool water.

"Over there!" said Mom, pointing to some brown mushrooms. "Those mushrooms are dangerous. Even though at first they look like mushrooms we could eat, when you look closer, you can see that they have white gills underneath the cap and a thick white ring around their stem. It is safest to always be observant."

I then listened as she told Otto and me things that were not okay for others to do to us.

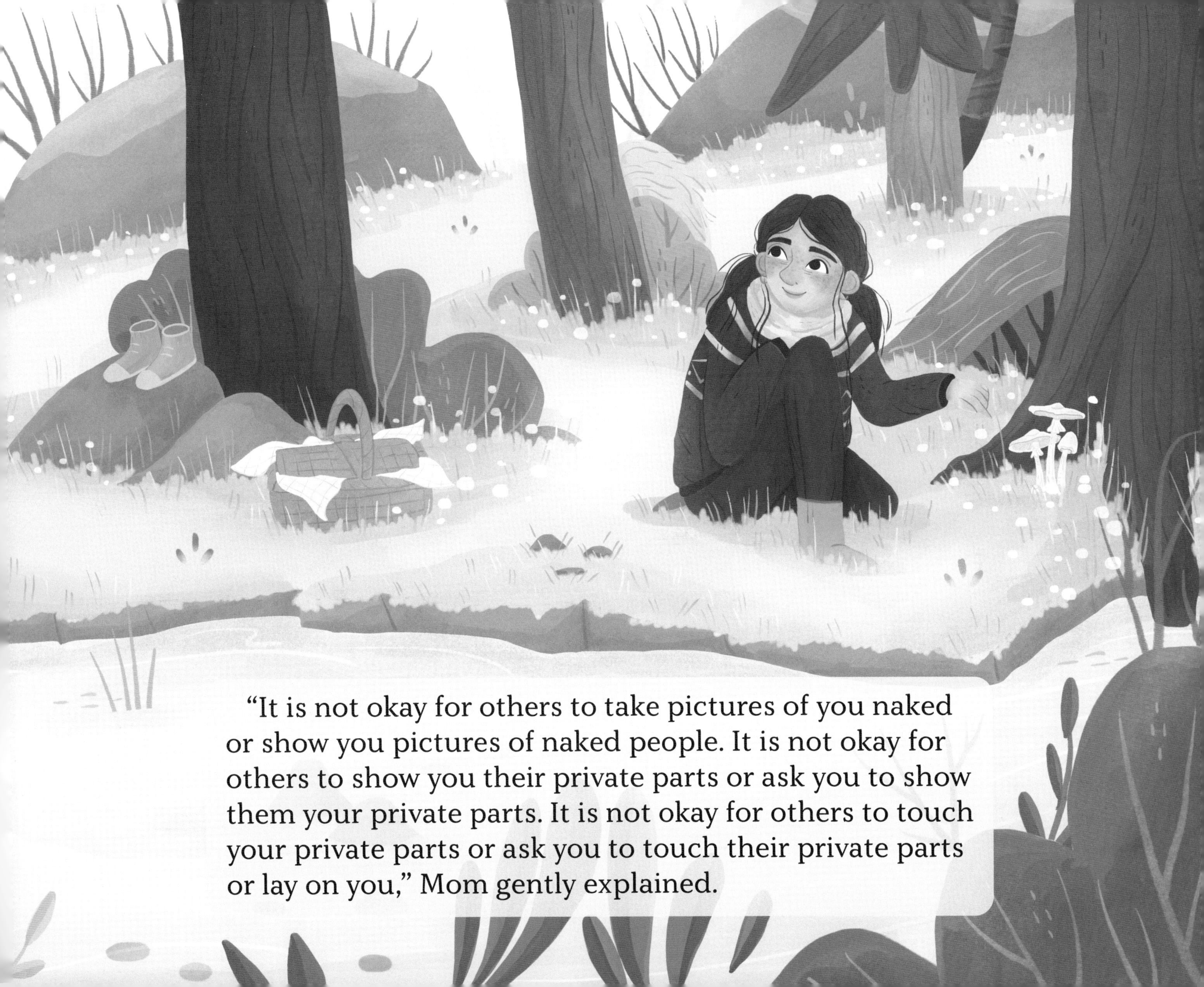

"It is not okay for others to take pictures of you naked or show you pictures of naked people. It is not okay for others to show you their private parts or ask you to show them your private parts. It is not okay for others to touch your private parts or ask you to touch their private parts or lay on you," Mom gently explained.

"What do we do if those things happen?" asked Otto.

"Say no!" Mom said. "Run away, and scream if you can."

She turned to Otto. "You remember how I tell you to not kick?"

Otto nodded his head.

"Well, if someone is trying to make you do the things I just told you about, it is okay to scratch, bite, or kick. Anything that helps you get away from them."

Otto stood up and started kicking. "Like this?" he said.

Mom and I laughed.

We put our shoes back on and walked across a bridge. I breathed in the smell of pine needles and wildflowers.

"One thing you should know," Mom said, "is that if someone ever does those things to you, it is never, ever, ever your fault, no matter what."

“You should always tell a trusted adult if someone does those things or even talks about those things,” Mom said, looking at us softly. “No matter what, you will never, ever be in trouble.”

Mom stopped and showed us weird, huge mushrooms at the bottom of a tree.

"These are called hen-of-the-woods! They are safe, and I love to make soup with them."

We put them in Mom's basket.

We came out of the forest at the top of a huge grassy hill. Otto and I rolled down while Mom laughed and ran down the hill.

We sat at the bottom of the hill to catch our breath.

“Sometimes people will threaten you to try to get you not to talk about bad things they said or did. They might say they will hurt you or your family or that you will be in trouble. But that is not true. Always tell a trusted adult.”

"Like a parent?" I asked.

"Yes," said Mom. "Or a teacher, or a relative, or a person at church."

We started our trip back home. The clouds floated above us as we walked.

As we walked home, Mom taught us more about safe and unsafe mushrooms, pointing out the unsafe ones and placing more edible ones in her basket.

Feeling content, I thought, "Bad things happen in our world, so I am glad Mom taught me about them. But, although there are bad things, there are so many beautiful things, and God will always be there to help us through bad things."

And what do you think we ate that night? Mushroom soup!

LEARN MORE ABOUT SAFETY WITH THE GOOD AND THE BEAUTIFUL LIBRARY

GOODANDBEAUTIFUL.COM

LEARN MORE ABOUT SAFETY WITH
THE GOOD AND THE BEAUTIFUL LIBRARY

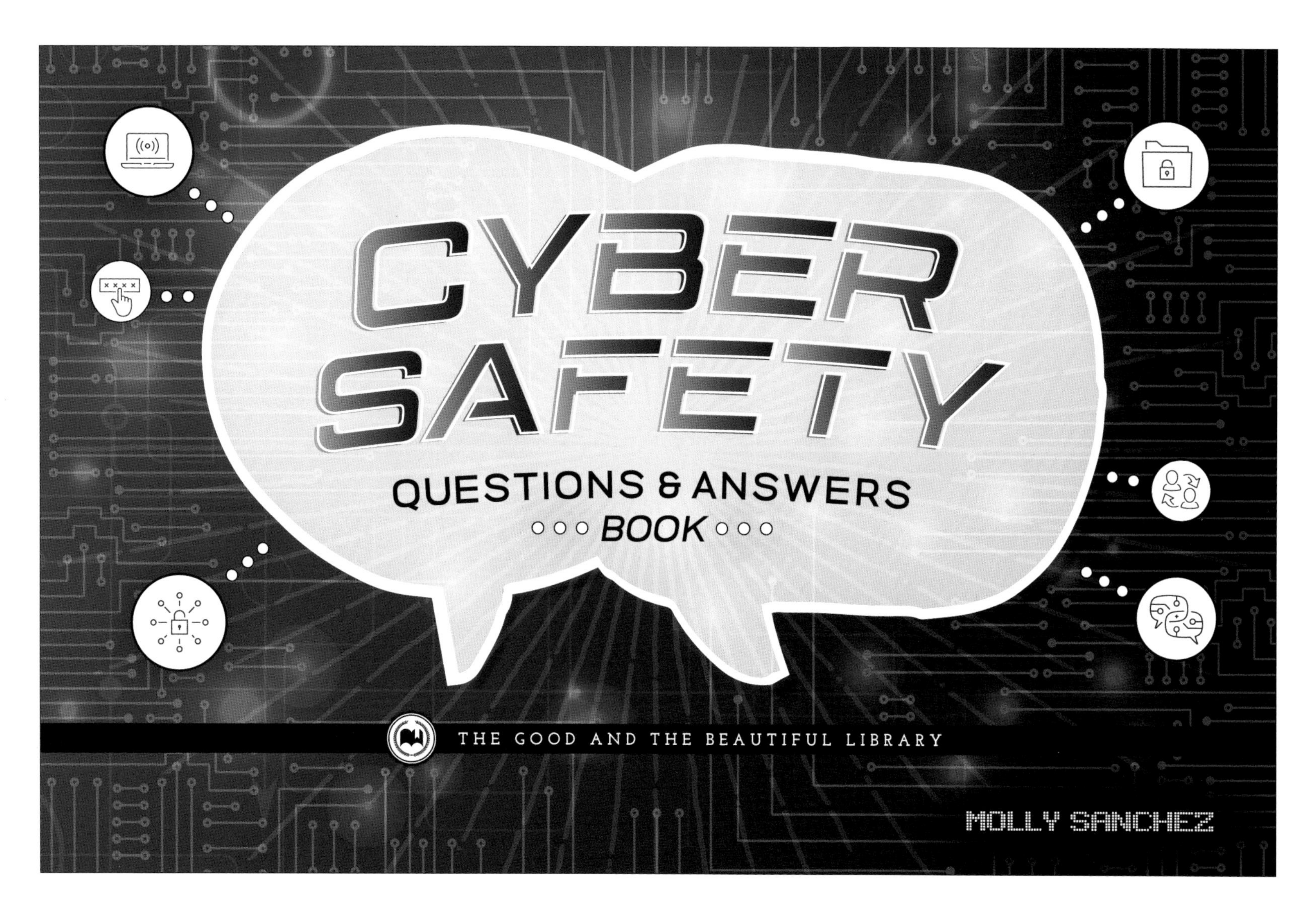

GOODANDBEAUTIFUL.COM